Little Scientists **BIG** Questions **?**

How Does an Engineer Use Science?

By **Ruth Owen**

Design by **Alix Wood**

What is an engineer?

3

Engineers invent, design and build the things that people use every day.

What does design mean?

It means to think up how something will look and work.

The things engineers build solve problems in our world.

To do their work, engineers use lots of **science**.

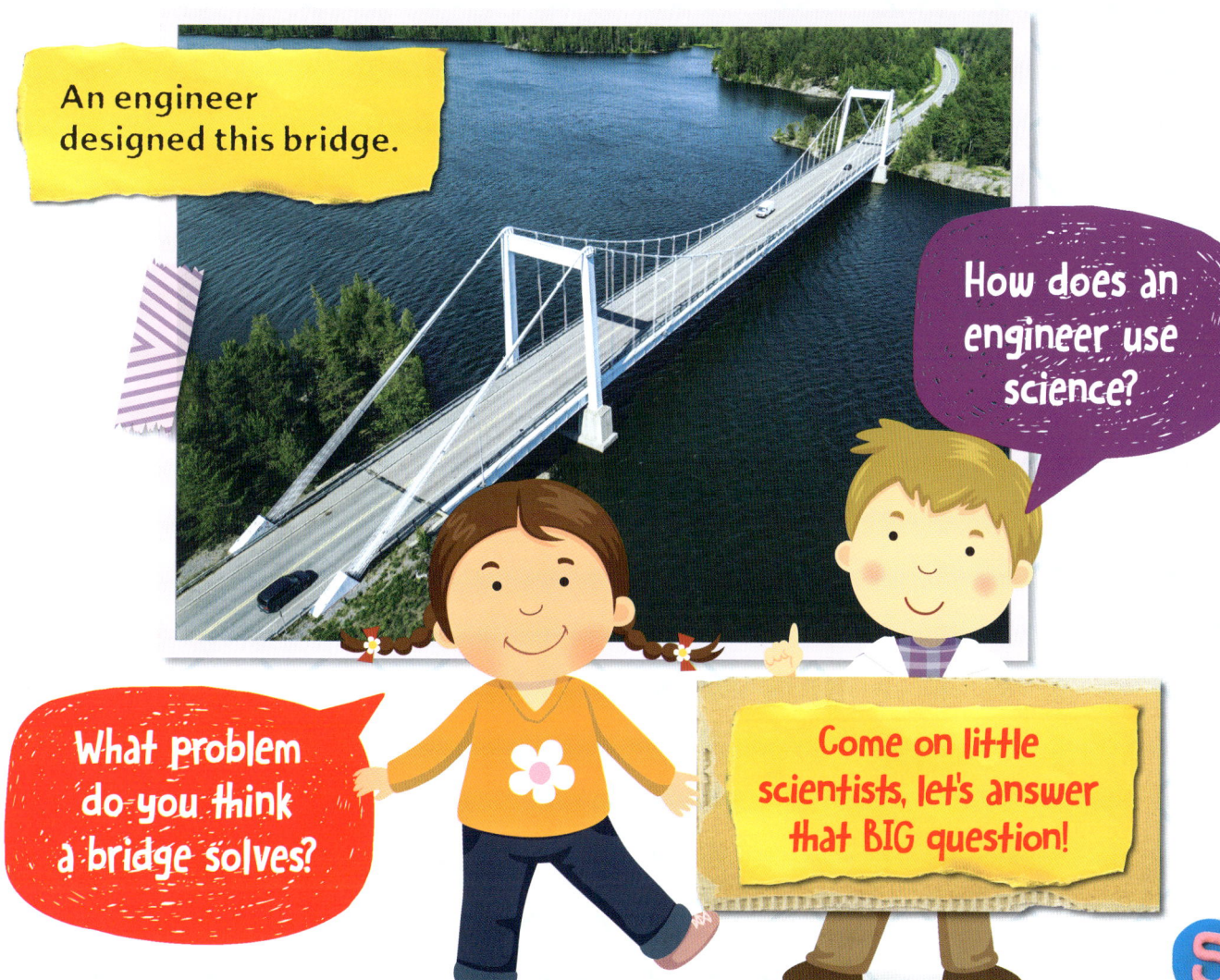

An engineer designed this bridge.

How does an engineer use science?

What problem do you think a bridge solves?

Come on little scientists, let's answer that BIG question!

There are lots of different types of engineers.

Some engineers design and build skyscrapers, roads and bridges.

Some engineers build engines and other machines.

A plane engine

Robot and computer engineers design and build robots.

These robots work in a big building called a warehouse.

Things for sale

Each robot has a computer inside!

Robot

The robots are carrying books, games and other things for sale.

People buy these things online.

Get ready for some BIG science!

To invent something new, an engineer works step by step.

Let's work with a space robot engineer!

STEP 1 :
What is the problem?

space robot engineer

Mars

How can we explore the surface of Mars?

They use space science, robot spacecraft and computers.

This robot spacecraft flies around Mars and sends information back to Earth.

It is freezing cold on Mars.

There is no air on Mars.

Mars is millions of miles from Earth.

It is too dangerous for scientists to go there.

space robot engineer

9

STEP 3 :
Design something that solves the problem.

A robot rover could go to Mars and explore the planet's surface.

A robot doesn't need air or water. It won't care that it's dangerous on Mars.

Engineers draw their designs on paper.

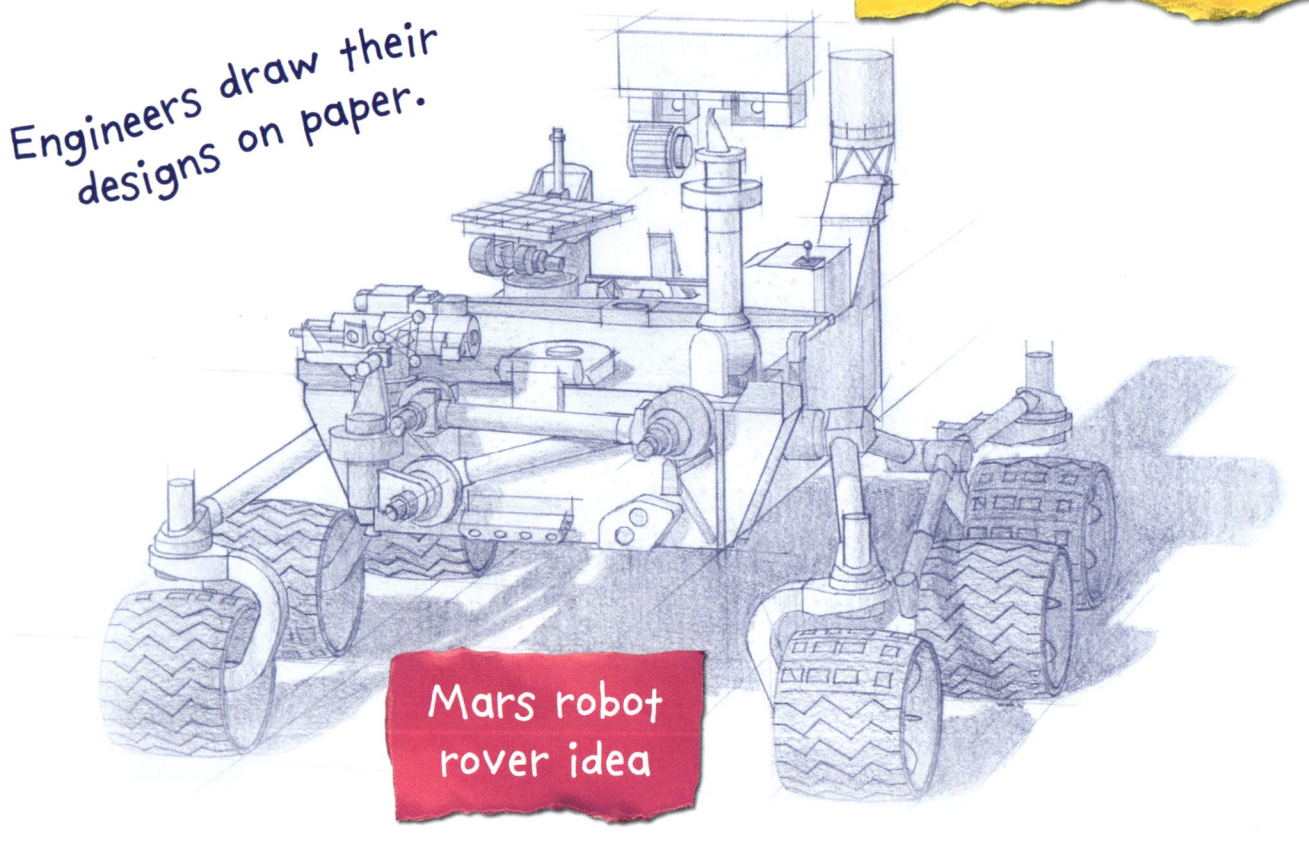

Mars robot rover idea

Sometimes they design them on a computer.

STEP 4 :
Build a prototype.

Next, an engineer uses science and tools to build a prototype, or model, of their idea.

They use the model to test if their design will work.

space robot engineer

Prototype

STEP 5 :
Test, test and test again!

Once an engineer builds their idea, they test it again and again.

These engineers are testing if the robot can drive over bumpy ground on Mars.

Robot rover

Engineers

This robot passed its tests. It's ready to go to Mars!

space robot engineer

STEP 6 : Try again!

Sometimes an idea doesn't work and an engineer must try again.

But don't worry! This can make the idea better next time.

Oh dear! Our robot needs more work.

This is a robot called *Perseverance*. It is exploring Mars.

Engineers invented a machine called a 3D printer.

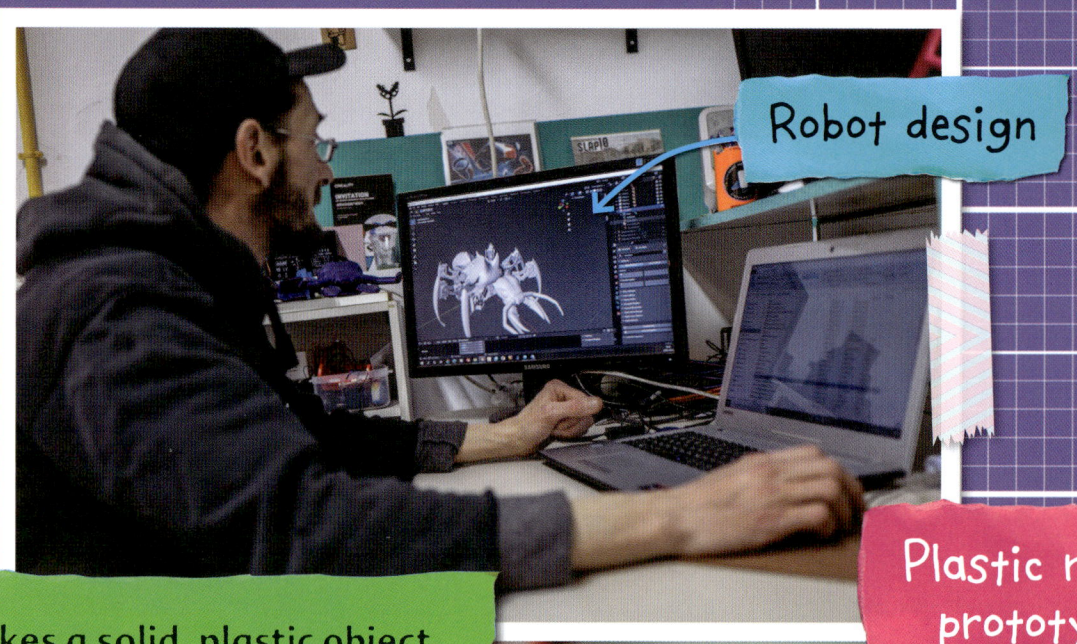

Robot design

Plastic robot prototype

It makes a solid, plastic object from a picture on a computer.

Engineers use 3D printing to make prototype models.

How does a 3D printer work?

Computer

Solid object

Liquid plastic

Printer

1: The engineer draws an object on a computer.

2: The printer copies the drawing.

3: The printer makes the object from thin layers of hot, liquid plastic.

4: The plastic cools and goes solid.

Some engineers wanted to help children with limb difference.

A child may be missing a part of their arm because they had an accident.

Some children are born with limb difference.

Plastic arm and hand

The engineers invented plastic arms and hands that children can wear and use.

The arms and hands are called prosthetic limbs. The engineers used lots of science facts about the human body to think up their designs.

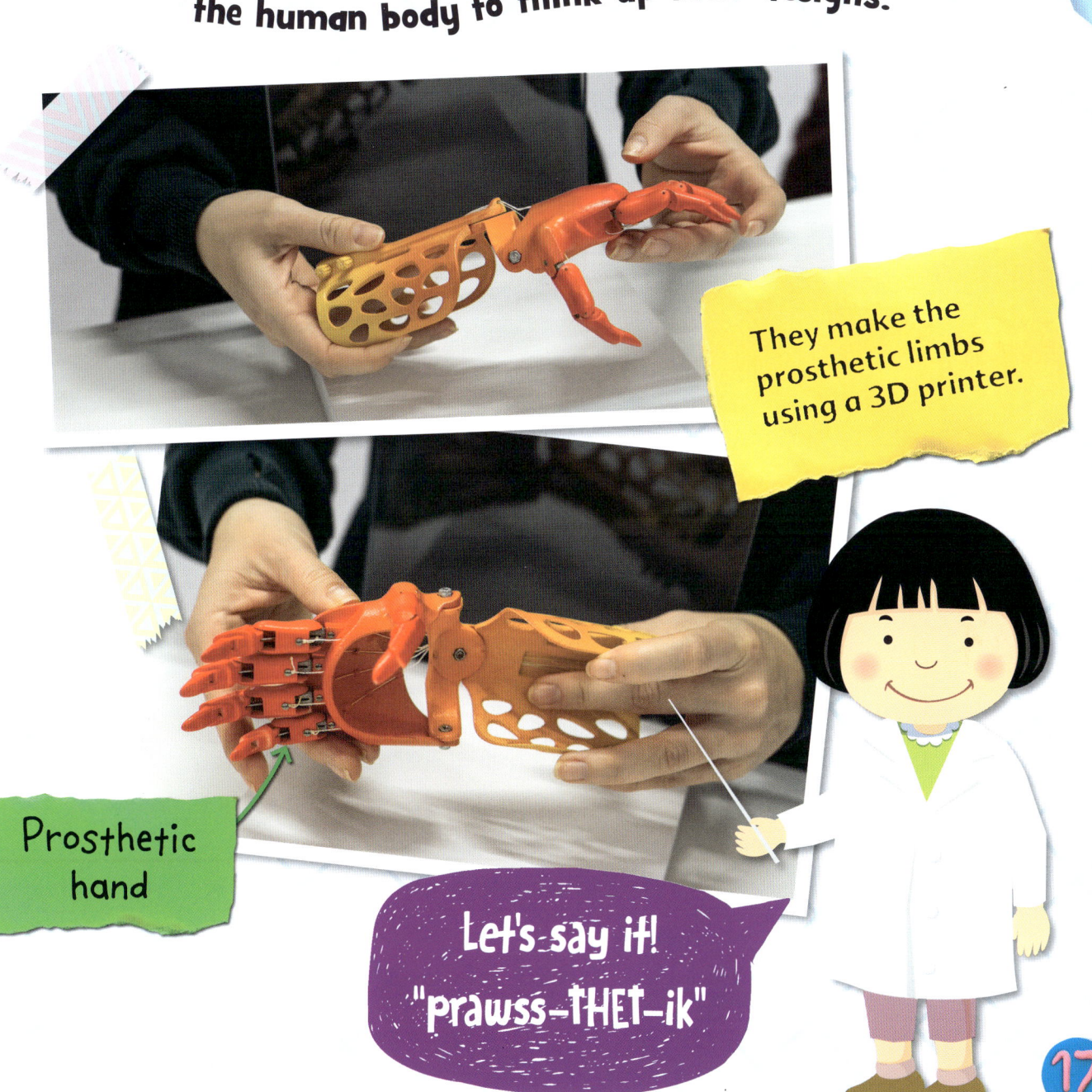

They make the prosthetic limbs using a 3D printer.

Prosthetic hand

Let's say it! "prawss-THET-ik"

Some engineers design things to help our planet.

They design and build wind turbines that make eco-friendly electricity.

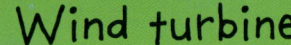

Wind turbine

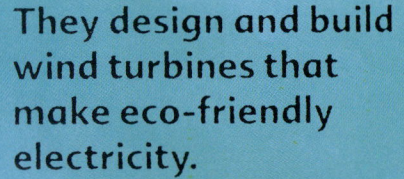

Engineer

The blades of a wind turbine turn a machine that makes the electricity.

Prototype

A wind turbine engineer must know the science of how wind blows.

Blade

Engineer

Blade

Then they can design blades that catch lots of wind to make lots of electricity.

A rollercoaster engineer must know all about forces and energy.

A motor pulls the rollercoaster cars to the top of the first hill.

Rollercoaster cars

Then a force called gravity pulls the cars down!

As the cars whizz down the big hill, energy is created.

The energy pushes the cars around the rest of the ride.

Engineers use science to design a fast and exciting ride.

21

Some engineers know all about electricity and machines.

They come to our homes to fix ovens, dishwashers and other machines.

I need help. Call an engineer!

My Science Words

eco-friendly
Something that is good for our planet and doesn't cause damage or pollution.

gravity
The force that pulls everything on Earth down towards the ground.

invent
To think up and create something new.

prototype
A model or first version of something new that can be used for testing.